Igor Holanda

Measurement of grounding resistance and ground potentials

Igor Holanda

Measurement of grounding resistance and ground potentials

Application example

ScienciaScripts

Imprint

Any brand names and product names mentioned in this book are subject to trademark, brand or patent protection and are trademarks or registered trademarks of their respective holders. The use of brand names, product names, common names, trade names, product descriptions etc. even without a particular marking in this work is in no way to be construed to mean that such names may be regarded as unrestricted in respect of trademark and brand protection legislation and could thus be used by anyone.

Cover image: www.ingimage.com

This book is a translation from the original published under ISBN 978-613-9-63708-9.

Publisher:
Sciencia Scripts
is a trademark of
Dodo Books Indian Ocean Ltd. and OmniScriptum S.R.L publishing group

120 High Road, East Finchley, London, N2 9ED, United Kingdom
Str. Armeneasca 28/1, office 1, Chisinau MD-2012, Republic of Moldova, Europe
Printed at: see last page
ISBN: 978-620-7-68903-3

DEDICATORY

I dedicate this work to those who are my foundation, my parents Valdemir and Antonia and my sister in memory, Carolina, who are everything to me and who gave me a lot of support in the most difficult moments of my life and the strength to carry on and never let me give up. And thank you very much for everything you've done throughout my life, I wish I'd been worthy of the effort you've put in.

ACKNOWLEDGEMENT

I would firstly like to thank GOD for his strength and courage throughout this journey, to whom I owe everything I am.

I would especially like to thank my parents who have always supported me in achieving my degree, always tirelessly, because they are the example for my life. I dedicate this work entirely to them

To my sister who always encouraged and supported me.

To all my family members for their help, support and encouragement. Especially my aunt Verônica, for her effort, trust and help.

To all my friends who supported me, corrected me and for understanding my disappearance, but who were always around, willing to help me, listening to my anxieties and sharing happy moments.

I'd like to thank my advisor Marcus and Luiz for the opportunity to learn every day and for helping me grow professionally and personally.

I'm grateful for all the difficulties I faced, they were worthy adversaries and made my victories much more palatable.

"There are men who fight one day and are good. There are others who fight for a year and are better. There are those who fight for many years and are very good. But there are those who fight all their lives. Those are the indispensable ones."

BertoltBrecht.

SUMMARY

This work will show the basic concepts of an electrical earthing grid, measuring earthing resistance and potentials on the surface of the ground in the grid. The procedures currently used for measurements will be shown, which are based on the IEEE Std. 81-2012 and ABNT NBR 15749-2009 standards. Initially, the definition and objectives of an earthing system will be presented, with the aim of better understanding the measurements of this system, through earthing resistance, touch voltage and step voltage. To complement the work, some of the most suitable methods for measuring resistance and potentials on the surface of an earthing system were explained. Finally, the aim of this work is to validate an existing method by means of a case study for a substation grounding grid with a supply voltage of 69kV, which seeks to provide a safe assessment of a system. The study is presented, containing real data obtained from carrying out the tests that will be mentioned.

Keywords: Measurements; Earthing resistance; Step and touch voltage.

SUMMARY

CHAPTER 1

INTRODUCTION

Nowadays, buildings require some form of earthing, whether for protection in the event of a fault in the electrical system, to dissipate static electricity or to protect against atmospheric discharges and manoeuvring surges. In order to perform satisfactorily, to be sufficiently safe against accidents and to protect equipment, every electrical installation must have a correctly dimensioned earthing system.

The main objectives of the earthing system are:
> Security protection action;
> Protection of installations against atmospheric discharges or manoeuvring surges;
> Protecting people from coming into contact with metal parts of the energised installation and standardising the potential of the mesh throughout the project area, preventing injuries in the event of a system failure.
> Provide an electrical path for normal and abnormal currents to flow through the system;
> Ensure the integrity of the installation's equipment.

The aim of this work is to clarify and verify the performance of the earthing grid design, with regard to maximum touch and step potentials and earthing resistance, as a function of the dimensions of the grid designed and the short circuit levels considered for this substation. It is necessary to check the levels of the earthing system by measuring earthing resistance and potentials on the ground surface, using the methods mentioned in this work.

The ABNT NBR 15749:2009 standard (Measurement of grounding resistance and potentials on the ground surface in grounding systems) establishes the criteria and methods for measuring the resistance of grounding systems and potentials on the ground surface, as well as defining the general characteristics of the equipment that can be used in measurements and the concepts for evaluating the results.

In the event of energisation in open substation networks, the circulation of currents through the ground generates potential gradients between the feet of an individual in the area (step voltage) or between their feet and a grounded metal point that they are touching (touch voltage). Therefore, these potential calculations are not compulsory for all meshes in places where people or animals may be present at dangerous potentials. It should be emphasised that the correct sizing of the mesh is to predict potentials and then confirm, through specific measurements, their possible changes in the workplace. The transit of people must be carried out with the utmost care and adequate time, as it directly concerns safety, so for these future measurements we must choose the potential profiles based on the *layout of* the installation.

This paper presents a case study with the aim of analysing the results obtained during an inspection of the earthing system of an overhead substation in a textile industry with an installed power of 5/6.25 MVA and a voltage supply of 69kV. After all the measurements taken, it was found that the grounding resistance and touch and step voltage values were satisfactory, and the assessment criteria of Coelce standard NT-004/2011 R05 (High voltage electricity supply - 69 kV) and IEEE Std 80-2000: "IEEE *Guide for Safety in AC Substation Grounding*" were adopted.

A study has been carried out on the results to check whether the measured values are lower than those stipulated by the standards.

The standards described above set limits for voltage levels and the maximum value of the earthing resistance in relation to a given reference voltage. During the measurement period, there were no readings outside the range considered appropriate by the standards.

1.1 Work organisation

Chapter 2 covers the theoretical background to the earthing system, the characteristics of the soil, which are described in the basic concepts of soil resistivity and the factors that influence its value, the earthing resistance and a generic way of calculating it. The potentials of the earthing system are also presented, as well as a procedure for calculating touch and step voltages.

Chapter 3 presents the procedures for measuring grounding resistance and potentials on the surface of the ground, but only a few of the existing methods were covered, which were necessary in the scope of this work.

Chapter 4 presents the results of the inspection of a grounding grid in a substation with a supply voltage of 69 kV, through a case study. Only the measured values were checked and compared with the maximum values established by the standards.

It should be emphasised that there was no initial sizing of the mesh, and this data was sent by the designer who sized the system. In this study, only the grounding resistance and potential values were measured, verifying the actual values against the standards in force.

Chapter 5 shows the conclusions drawn from this work.

Finally, the drawing of the earthing mesh and its dimensions will be presented in the annex on an A1 sheet.

CHAPTER 2

THEORETICAL FOUNDATIONS

2.1 Introduction to earthing systems

The definition of an earthing system according to ANBT NBR 15749 (2009) is: "The set of all earthing electrodes and conductors, whether interconnected or not, as well as metal parts that act directly or indirectly with the earthing function, such as: towers and gantries, building reinforcements, metal cable covers, pipes and the like".

The objectives of an earthing system are:

> Actuation of the safety protection, allowing the circulation of the fault current, with higher values for faster and more efficient protection;

> Protection of installations against atmospheric discharges and/or manoeuvring surges, providing current flow to earth;

> Protecting people from contact with metal parts of the energised installation and standardising the potential of the mesh throughout the project area, preventing injuries in the event of a system failure.

> Provide an electrical path for normal and abnormal currents to flow into the system and limit overvoltages in the event of faults;

> To allow the accumulation of static electricity to flow into the ground;

> Ensure low grounding resistance;

> Ensure the integrity of the installation's equipment.

An efficient earthing system depends primarily on the method of distribution in the structures, the electrode system used, the resistivity of the soil, the efficiency of all existing connections and, finally, the materials used in the earthing mesh. The various types of earthing systems must be realised with the aim of guaranteeing the best connection with the earth, ensuring a system that is adequately dimensioned for the conditions of each project.

The main elements of the earth grid, seen in Figure 1, are basically made up of:

- Earthing electrode or vertical electrode;
- Earthing conductor;
- Connection or splice;
- Protective conductor.

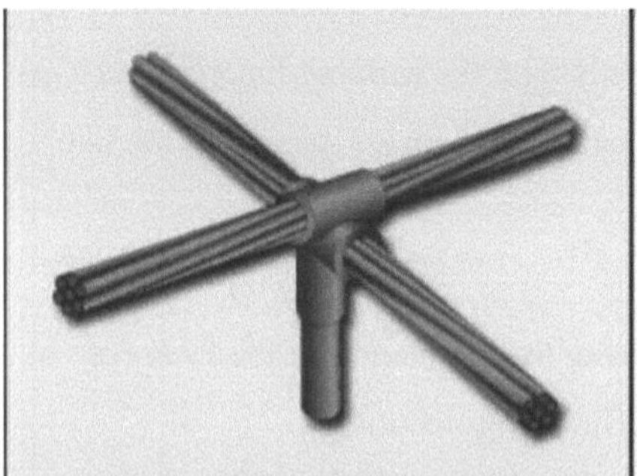

Figure 1: Elements of an earthing mesh.
Source: (Leite, 2007)

The point in the system that you want to connect to ground can be of various types, depending on your application. It can be a track on a printed circuit board, the housing of a motor or computer, or the neutral of an electrical system.

Earthing electrodes can have various configurations, but basically they are made up of any metal body buried in the ground, such as a bar of galvanised or copper-plated steel material, galvanised iron angles, hydraulic systems or meshes of copper conductors in reticulation. A

he geometric arrangement of the electrodes in the soil depends on their application. The most common is the use of vertical rods, used mainly when the deeper layers of the soil have lower resistivity, and horizontal elements that are buried at a certain depth are used to control the potential gradient on the soil surface (VISACRO FILHO, 2002).

2.1.1 Soil resistivity

The soil is the medium in which the earthing system will be located, so its electrical properties will be determining factors for sizing. As we want to conduct an electric current to the ground, the relevant property will be resistivity, which will indicate a resistance to the passage of electric current.

To design an earthing system, the first step is to have prior knowledge of the soil's characteristics. To obtain the soil's resistivity values, it will be necessary to take measurements with a terrometer-type instrument.

Soil resistivity can be defined as the resistance between opposite sides of the volume of a cube of homogeneous, isotropic soil whose edge measures 1 metre. Its unit is "$\Omega.m$". Illustrated in Figure 2.

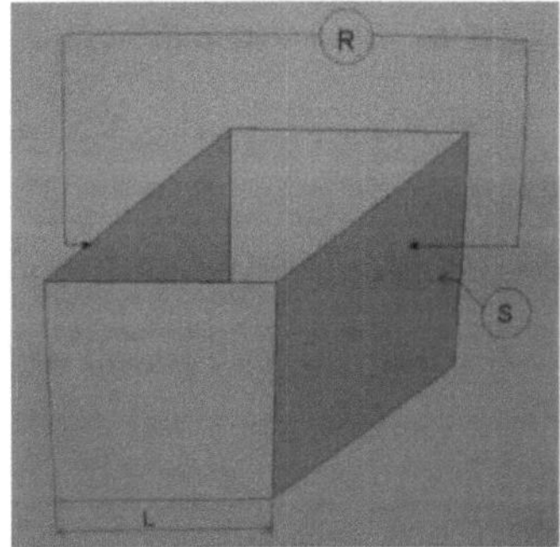

Figure 2: Soil resistivity cube.
Source: (Leite, 2007)

Several factors influence soil resistivity, as shown below:

-Type of   soil;

-Soils   made up of stratified layers with different depths and materials.

   -Humidity content;

-Compaction   and pressure;

   -Chemical composition of   the salts dissolved in the retained water.

The various combinations above result in different soil characteristics and different resistivity values.

The resistivity value of the soil is influenced by its type, composition, temperature and water retention capacity. Soil is a heterogeneous element, so its value varies from one direction to the other, depending on the composition of its material, the depth of its layers and the age of its geological formation.When designing an earthing grid, the values in Table 1 should never be used, as these values only illustrate the resistivity of soils of different natures.

Table 1: Type of soil and its resistivity.

Soil type	Resistivity (Ω.m)
Mud	5 a 100
Garden soil with 50% moisture	140
Garden soil with 20% moisture	480
Dry clay	1500 a 5000
Clay with 20% moisture	330
Clay with 40% moisture	80
Wet sand	1300
Dry sand	3000 a 8000
Compact limestone	1000 a 5000
Granite	1500 a 10000

Source: (KINDERMANN, et al., 2011)

2.1.1.1 Influence of humidity

The resistivity of the soil changes with the variation in humidity. The variations occur because the conduction of electrical charges in the soil is predominantly ionic. Soil containing a higher percentage of moisture favours the dissolution of the salts present, forming an electrolytic medium favourable to the passage of ionic currents.

Thus, a specific soil with a moisture concentration of "x" shows a large variation in its resistivity. Table 2 shows the variation in soil resistivity in relation to humidity.

Table 2: Resistivity of a soil with moisture concentration.

Moisture content (% by weight)	Resistivity (Í2.m)
0	10.000.000
2,5	1.500
5	430
10	185
15	105
20	63
30	42

Source: (KINDERMANN, et al., 2011)

According to Kindermann (2011, p.3): "It is therefore concluded that the value of soil resistivity follows the periods of drought and rainfall in a region. Earthing improves its quality with wet soil, and worsens in the dry season." It is therefore recommended that the earthing system be dimensioned for the driest time of the year.

Some actions are taken because of the influence of moisture on soil resistivity. For example: in arid soils, it is sometimes necessary to use deep rods in order to reach layers with higher humidities.

If the soil has deep layers that are less resistive than the surface layers, this can help reduce evaporation. Another way is the presence of gravel in substations, favouring the reduction of evaporation.

2.1.1.2 Influence of chemical composition

The influence of chemical composition depends on the presence and quantity of soluble salts and acids that are normally found aggregated in the soil, which will predominantly influence its resistivity value. As proposed by Mamede Filho (2012 p.401):

> It is well known that when it is necessary to reduce the resistance of a given earthing system, chemical products are appropriately added to the soil surrounding the earthing electrode. There are various chemical products based on a mixture of salts which, when combined with each other and in the presence of water, form GEL, a commercially used product that is highly effective in reducing soil resistivity. These compounds have the following characteristics:
> - They are hygroscopic   ;
> - They give   the soil   chemical stability;
> - Non-corrosive;
> - They are not   attacked by acids;
> - They are insoluble in the presence of water;
> - It lasts a long time (usually 5 to 6 years).

2.1.1.3 Influence of temperature

Temperature has an effect on soil resistivity, which can reach high values at extremely low temperatures. To avoid such variations, the electrodes should be at a depth where there are small variations in temperature. So, keeping all the other characteristics fixed, and varying only the temperature, the typical effect on soil resistivity is shown in table 3.

Table 3: Resistivity of a soil with temperature variation.

Temperature (°C)	Resistivity Ω.m)
20	72
10	99
0 (water)	138
0 (ice)	300
-5	790
-10	3.330

Source: (KINDERMANN, et al., 2011)

Generalising, the performance of a given soil subjected to temperature variation can be expressed by the curve shown in Figure 3.

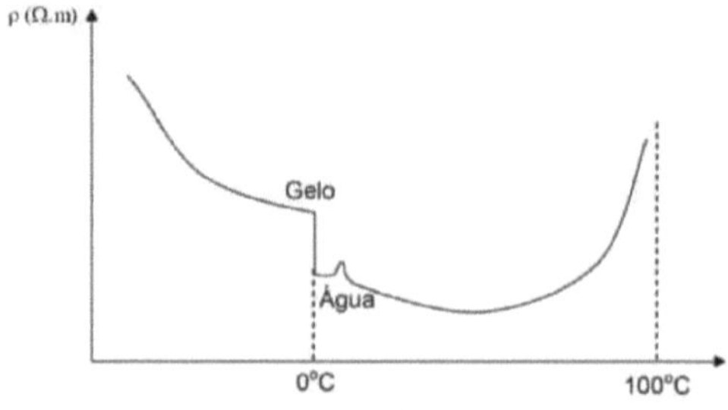

Figure 3: Effect of temperature on soil resistivity.
Source: (KINDERMANN, et al., 2011)

2.1.1.4 Apparent soil resistivity ("ρ_a ")

The apparent resistivity of the soil is the resistivity seen by a specific earthing system. For example, a homogeneous soil can have different resistivity values or the system can interact differently on the same resistance parameter.

The value of the electrical resistance of an earthing system depends on two essential factors: the apparent resistivity of the soil and the geometry of the mesh, including the number and spacing of cables and rods adopted in the design.

2.2 Earthing resistance

The earthing system is influenced by frequency and can be characterised electromagnetically by means of an earthing impedance.

To assess the nature of earthing, it should be considered that in general, an earth connection has resistance, capacitance and inductance, each of which influences the ability to conduct current to earth. The impedance is the way in which the system sees the earthing. This impedance, called grounding impedance, can be conceptualised as the opposition offered by the ground to the injection of electric current into it, through the electrodes, and is expressed quantitatively through the relationship between the voltage applied to the ground and the resulting current (VISACRO FILHO, 2002).

To better illustrate the nature of this impedance, Figure 4 shows a simplified representation of earthing, using an equivalent circuit, and briefly explains the origin of its configuration.

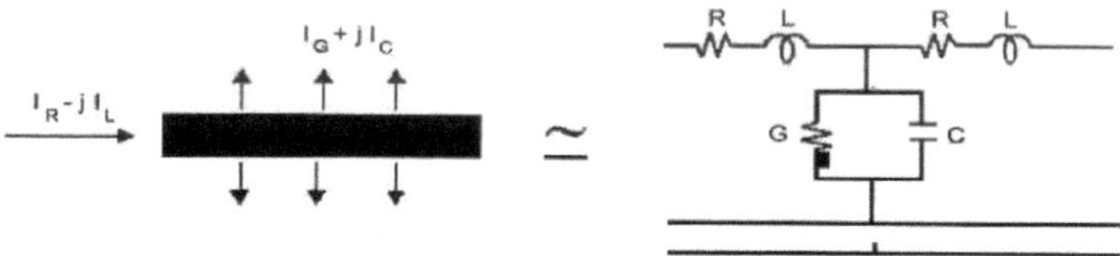

Figure 4: Current components in the ground.
Source: (VISACRO FILHO, 2002).

If a limited portion of the electrode is considered, it can be seen that the current dissipated into the ground is made up of the four components shown in figure 4. The current that is injected into the electrode is partially dissipated into the ground and partially transferred to the remaining length of the electrode. As far as the latter is concerned, in longitudinal currents there are losses internal to the conductor and a magnetic field is established in the region around the conductor due to the passage of this current, both inside and outside the conductor. On the other hand, the electric field in the ground (medium of resistivity "ρ" and permittivity "ε") determines the flow of conductive and capacitive currents in the medium. The relationship between these currents does not depend on the geometry of the electrodes, but only on the ratio "σ/ωε", where "σ" refers to the conductivity of the soil and "ω" to the angular frequency. In terms of an equivalent circuit, the corresponding energies can be calculated using a resistance and an inductance in series.

In many applications, this does not refer to the grounding impedance, but to its resistance. This is because when analysing low frequency phenomena (industrial frequency), the reactive effects are very small and can be disregarded. In this case, the equivalent circuit for earthing is reduced to a series of coupled conductances. Thus, at low frequencies, the earthing system can be characterised electromagnetically by means of an earthing resistance, i.e. for sizing short circuit earthing systems, capacitive and inductive effects are disregarded, as seen in Figure 5.

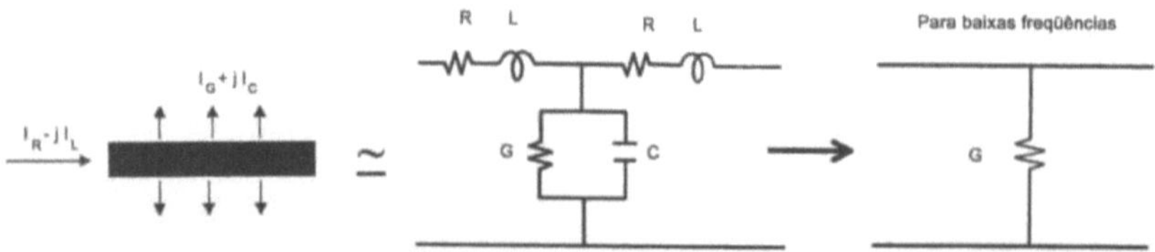

Figure 5: Equivalent grounding circuit under low frequency conditions.
Source: (VISACRO FILHO, 2002)

According to Mamede (2010, p.397):

> In an earthing system, the effect of three resistances is considered to be earth resistance:
>
> • The relative resistance of the connections between the earth electrodes (rods and cables);
>
> • The relative contact resistance between the earth electrodes and the ground surface around them;
>
> • The resistance relative to the ground in the vicinity of the earth electrodes is also called dispersion resistance.

2.2.1 Interference zones

By inserting rods in parallel, we reduce the value of the grounding system's resistance, but for the purposes of calculating it, the parallel rods are not just the parallelism of the electrical resistance. This is because there is interference in the areas where equipotential surfaces act, where the passage of electric current in the ground is blocked, increasing the system's resistance. This is because the effective current dispersion area of each rod becomes smaller, and consequently the resistance of each rod within the set increases. Because of this, we must use considerable spacing between the rods in order to reduce this effect, as shown in Figure 6.

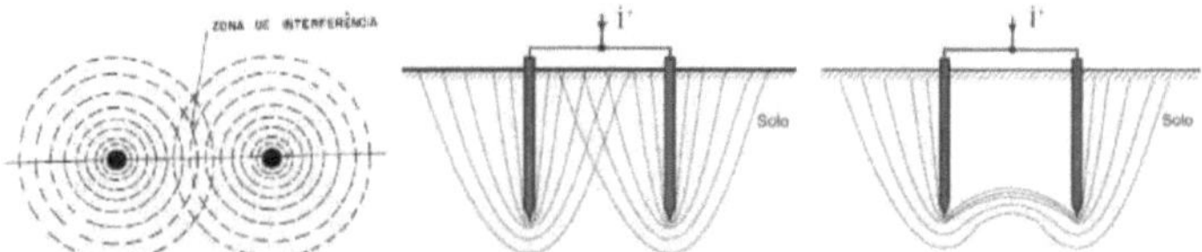

Figure 6: Interference zone in equipotential lines.
Source: (KINDERMANN, et al., 2011)

2.2.1 Resistance of the equivalent earth mesh considering parallelism of buried horizontal cables

The formula for calculating the resistance of an earthing system made up of buried horizontal conductors forming a grid is given by the IEEE std 80 (2000) standard.

$$R_1 = \frac{\rho}{\pi.L_c}.\left[\ln\left(\frac{2.L_c}{\sqrt{d_1.2h}}\right) + K_1.\left(\frac{L_c}{\sqrt{S}}\right) - K_2\right] \tag{2.1}$$

Where:

P is the resistivity of the soil or medium in Ω.m.

L_c is the total length of the horizontal conductors in metres.

d1 is the equivalent radius of the conductor in m.

Cmé the length of the earthing mesh in m.

Lmis the width of the earthing mesh in m.

S is the area of the earthing mesh arrangement in m^2.

K1 and K2 are coefficients.

$$S = Cm.Lm \tag{2.2}$$

$$y = \frac{Cm}{Lm} = \frac{comprimento}{largura} \tag{2.3}$$

$$K1 = 1,4125 - 0,0425 \cdot y \tag{2.4}$$

$$K2 = 5,49 + 0,1443 \cdot y \tag{2.5}$$

2.2.2 Equivalent resistance of vertical earthing electrodes in parallel

To complement the calculations, there is the rod resistance portion for the earthing system.

$$R_2 = \frac{\rho}{2 \cdot \pi \cdot Nh \cdot Lht} \cdot \left[\ln\left(\frac{4 \cdot Lht}{0,0254 \cdot Rht}\right) - 1 + 2 \cdot K_1 \cdot \left(\frac{Lh}{\sqrt{S}}\right) \cdot (\sqrt{Nh} - 1)^2 \right] \tag{2.6}$$

Where:

Lh is the length of the earth rod in m

Nh is the number of ground rods in the system

Lht=Lh.Nh ; Total Length of All Ground Rods (m)

Rht = Radius of earthing rod (m)

Dht = Earth rod diameter (m).

$$Rht = \frac{Dht}{2} \tag{2.7}$$

2.2.3 Mutual resistance between horizontal conductors and vertical electrodes

To finalise the system calculation, the mutual resistance between the horizontal conductors and the vertical electrodes is taken into account.

$$R_m = \frac{\rho}{\pi \cdot L_c} \left[\ln\left(\frac{2 \cdot L_c}{Lh}\right) + K_1 \cdot \left(\frac{L_c}{\sqrt{S}}\right) - K_2 + 1 \right] \tag{2.8}$$

2.2.4 Total resistance of the earthing system

The typical format for dimensioning an earthing system is based on the IEEE Std 80 (2000) standard, which shows a wide variety of equations for different types and parts of earthing systems. According to the standard, these equations have certain restrictions, such as:

- They are only valid for homogeneous soils;
- They focus on simple geometries;
- They use irregularity coefficients to model the non-uniformity of the electric current in conductors.

The value that represents the total resistance of the earthing system is the combination of the resistances of the horizontal conductors in parallel and the vertical electrodes in parallel and the mutual resistance between the mesh elements.

$$R_t = \frac{R_1.R_2 - R_m^2}{R_1 + R_2 - 2.R_m} \qquad (2.9)$$

Where:

R1 is the equivalent resistance of the buried conductors;

R_2 is the equivalent resistance of the vertical electrodes;

R_m is the mutual resistance between the conductors and the electrodes.

2.3 Potentials in the Earthing System

In the event of a fault in the earthing system, currents circulate through the ground, generating potential gradients on the surface that will cause potential differences between the feet of a person in the area, called step voltage, or with danger

This is between your feet and some grounded metal point that you are touching, called the touch voltage, as shown in Figure 7.

When an electric current is injected into the earth, whether due to a single-phase fault in an installation or lightning strikes, the currents dispersed by the earthing system cause a potential difference to appear between the two:

-Points   on the ground surface (step tension);

-Grounded metal parts of   the installation and the ground (touch voltage);

-Circuits   that are somehow connected to the earthing system and points far from the ground surface or other distant earthing systems (transferred potential, control and communication circuits, lightning cables, etc.).

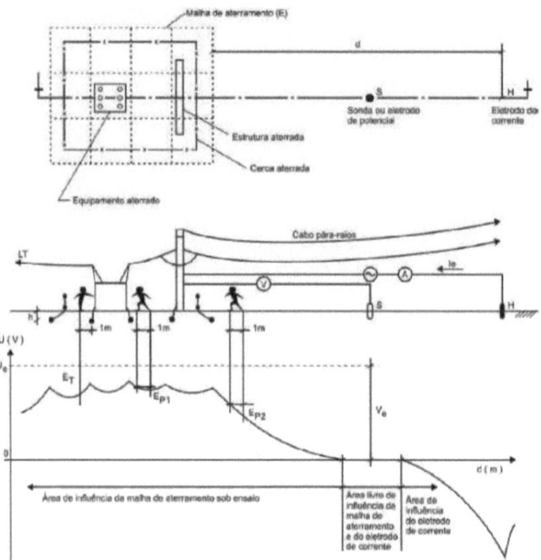

- **Figure 7**: Voltages that can appear in an installation.

Source: (ABNT NBR 15749, 2009)

Figure 8 shows, in general terms, situations involving a person and the earthing system during an electric current injection, showing a profile of the electric potential on the ground surface, indicating the various potentials produced.

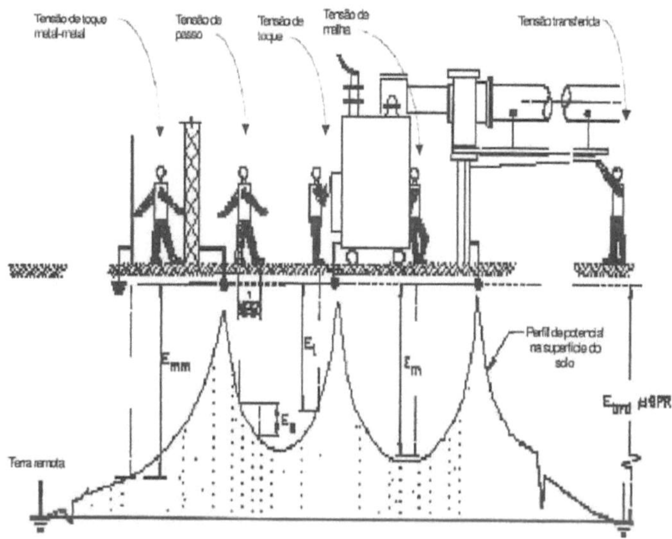

Figure 8: Typical situations where a person is subject to potential gradients.
Source: (TELLÓ, 2007)

2.3.1 Touch tension

This is the potential difference between a point on a grounded metal structure and a point on the ground surface separated by a horizontal distance equivalent to the normal reach of a person's arm. By definition, this distance is considered to be 1.0 m.

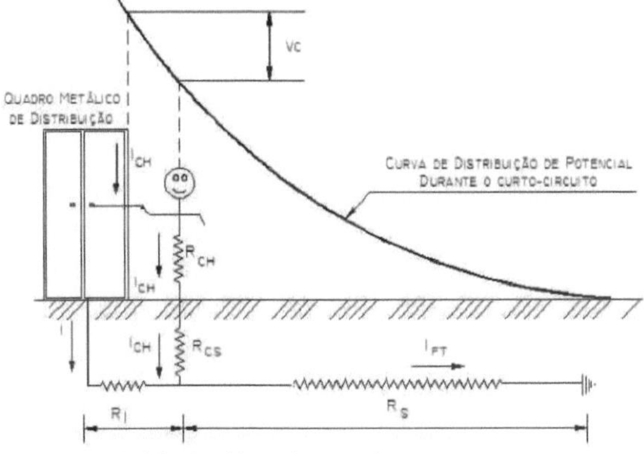

Figure 9: Individual subjected to touch stress.
Source: (MAMEDE, 2010).

Based on the IEEE 80:2000 standard, we take the values of the electrical resistance of the human body and the constants for the weight of the operator, which can be 50 or 70kg.

Determination of maximum touch tension:

$$E_{toque} = (Rch+Zth).Ib \tag{2.10}$$

$$E_{toque} = (Rch+1,5.Cs.\rho s).Ib \tag{2.11}$$

$$Ib = \frac{k}{\sqrt{Te}} \tag{2.12}$$

$$Cs = 1 - \frac{0,09.(1-\frac{\rho}{\rho s})}{2.hs+0,09} \tag{2.13}$$

$$E_{toque} = (1000+1,5.Cs.\rho s).\frac{k}{\sqrt{Te}} \tag{2.14}$$

Where:

ρ ; the resistivity of the soil;

Rch is the resistance of the human body;

Zth is the equivalent resistance of the system (Z Thevenin);

Te is the protection's operating time;

hs is the thickness of the material (gravel);

ρ s is the surface resistivity of the gravel;

Cs is the reduction factor due to surface densities and the type of surface material;

k: Empirical constant, related to the mass of the human being;

k = 0.116 for people with a mass of 50kg;

k = 0.157 for people with a mass of 70kg.

Determination of the touch voltage on the mesh:

$$E_m = \frac{\rho.K_m.K_i.Icft}{L_M} \tag{2.15}$$

$$K_m = \frac{1}{2.\pi}.\left[\ln\left[\frac{Dcp^2}{16.h.d}+\frac{(Dcp+2.h)^2}{8.Dcp.d}-\frac{h}{4.d}\right]+\frac{Kii}{Kh}.\ln\left[\frac{8}{\pi(2.n-1)}\right]\right] \tag{2.16}$$

$$n = n_a . n_b . n_c . n_d \tag{2.17}$$

$$K_{ii} = \frac{1}{(2.n)^{\frac{2}{n}}} \qquad\qquad (2.18)$$

$$K_h = \sqrt{1+\frac{h}{h_0}} \qquad\qquad (2.19)$$

$$K_i = 0{,}644+0{,}148.n \qquad\qquad (2.20)$$

Where:

K_m is the mesh coefficient;

K_i is the coefficient of irregularity;

Icft is the phase-to-ground short-circuit current;

L_M is the total length of the earthing conductors and rods;

Dcp is the distance between the conductors of the earthing grid;

n is the geometric coefficient;

d is the diameter of the conductor in m;

h_0 is the reference depth of the mesh;

h is the depth of the mesh.

2.3.2 Step voltage

This is the difference in potential between two points on the ground surface separated by the distance of a person's step, shown in Figure 10. By definition, this distance is considered to be 1.0 m.

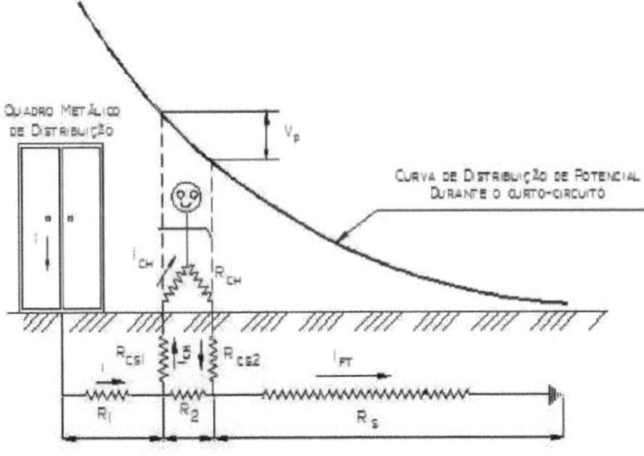

Figure 10: **Individual subjected to step tension.**
Source: (MAMEDE, 2010)

Based on the IEEE 80:2000 standard, we take the values of the electrical resistance of the human body and the constants for the weight of the operator, which can be 50 or 70kg.

Determination of the maximum step voltage:

$$E_{passo} = (Rch+Zth).Ib \tag{2.21}$$

$$E_{passo} = (Rch+6.Cs.\rho s).Ib \tag{2.22}$$

$$Ib = \frac{k}{\sqrt{Te}} \tag{2.23}$$

$$Cs = 1-\frac{0,09.(1-\frac{\rho}{\rho s})}{2.hs+0,09} \tag{2.24}$$

$$E_{passo} = (1000+6.Cs.\rho s).\frac{k}{\sqrt{Te}} \tag{2.25}$$

Where:

k: Empirical constant, related to the mass of the human being;

k = 0.116 for people with a mass of 50kg;

k = 0.157 for people with a mass of 70kg.

Determination of the step voltage in the loop:

$$E_s = \frac{\rho.K_s.K_i.Icft}{Ls} \tag{2.26}$$

$$K_s = \frac{1}{\pi}.\left[\frac{1}{2.h}+\frac{1}{Dcp+h}+\frac{1}{Dcp}(1-0,5^{n-2})\right] \tag{2.27}$$

Where:

K_s is the surface coefficient of conductors;

Ls is the percentage length of the conductors and earthing rods.

2.3.3 Correction of touch and step potentials

In order to reduce the effect of potentials, it is of fundamental importance in the substation area to cover the ground with a high resistivity material such as gravel, as this provides better insulation for foot contacts with the ground.

CHAPTER 3

METHODOLOGICAL PROCEDURES

To carry out resistance and potential measurements on the ground surface in earthing systems, we will adopt ABNT NBR 15749: 2009 (Measurement of grounding resistance and potentials on the ground surface in earthing systems), which establishes the criteria and methods for measuring the resistance of earthing systems and potentials on the ground surface, as well as the general characteristics of the equipment that can be used in the measurements and helps with the concepts of evaluating the results.

To determine values for research purposes, to check safety levels in installations in operation, or even for commissioning new installations, field tests are the most effective way of obtaining the values of the ohmic resistance of the earthing grid and the values of the step and touch potentials calculated in the project. Therefore, the resistance of the earthing grid associated with the potentials on the ground surface of an electrical installation are quantities to be measured, basically focusing on:

- Check the effectiveness of the mesh or earthing system in dissipating the electric current into the ground;

- Define changes to an existing earthing system in accordance with the requirements of the project and current standards;

- Check for possible touch and step voltages that pose a risk to living beings and equipment;

- Determine the potential rise of the earthing system in relation to the reference earth.

There are various methods for measuring earthing resistance and potentials on the ground surface, which will be mentioned later. Safety measures must be followed to minimise the risk of accidents related to dangerous potentials that may occur in the vicinity of earthing systems or grounded conductive structures. Certain technical and safety rules are recommended, such as:

- Wear shoes and gloves with an insulation level compatible with the maximum voltage values that may occur in the system under measurement;

- Avoid taking measurements under adverse weather conditions, given the possibility of lightning strikes;

- Prevent outsiders and animals from approaching the electrodes used in the measurement;

- Use equipment compatible with that specified in Annex C of ABNT NBR 15749:2009, in order to guarantee the safety and reliability of the results. Failure to comply with the safety procedures in Annex C makes it necessary to adopt additional safety measures, such as those used for work in energised areas.

3.1 Grounding resistance measurement

3.1.1 Potential drop method

This method basically consists of circulating a current through a circuit which is the earthing system we want to check the ohmic resistance value of, through the section of an auxiliary current electrode. At the same time, the voltage between the loop and the reference earth (remote earth) must be measured using a probe or auxiliary potential electrode. Figure 11 shows how the measurement is made. This method is recommended for measurements using specific equipment, such as a terrometer.

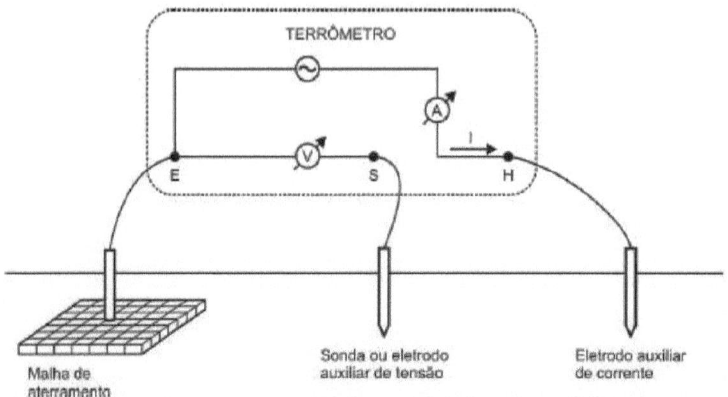

Figure 11: Potential drop method.

Caption
I Test current

SBorne to the probe or auxiliary potential electrode
HBorne for the auxiliary current electrode
E Terminal block for the earthing loop under measurement

Source: (ABNT NBR-15749, 2009)

The auxiliary current and voltage electrodes are each made up of one or more metal rods interconnected and firmly embedded in the ground, with the aim of guaranteeing the lowest grounding resistance of the whole.

In the measurement procedure, the probe or auxiliary potential electrode must be moved along a predefined direction, but generally in a straight line between the earthing mesh and the current electrode, which is measured from the periphery of the earthing system under test at regular measurement intervals equal to *COo OF* the distance "d" as shown in Figure 7. In this way, it is possible to acquire the resistance value at each position, obtaining a curve (Resistance x distance), as shown in Figure 12.

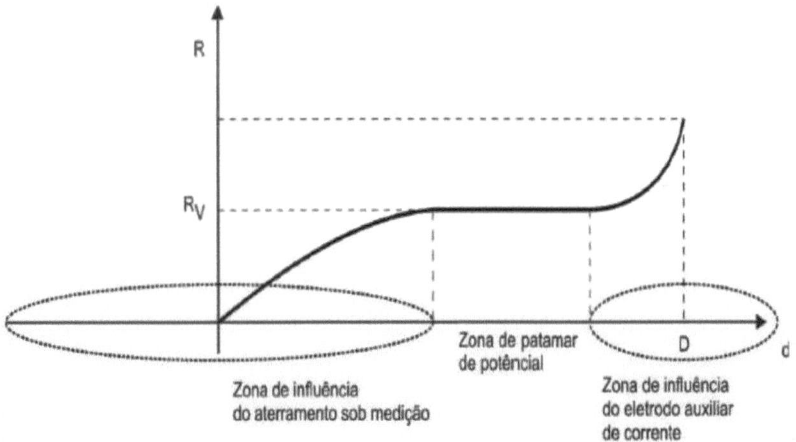

Figure 12: Theoretical characteristic curve of the grounding resistance of a point electrode.
Caption
R: Resistance obtained by varying the distance from the probe from distance d=D to d=0 (electrode to be measured)
RV: True ground value

Source: (ABNT NBR-15749, 2009)

This theoretical characteristic curve shown in Figure 12 has three

parts of the curve that are zones of mutual influence between the mesh under measurement, the

the auxiliary current electrode and the earth, and a zone called the "potential plateau",

where the true grounding value can be found.

Figure 13 shows typical grounding resistance curves as a function of the

spacing and relative position of the auxiliary potential and current electrodes in relation

to the grounding system under test.

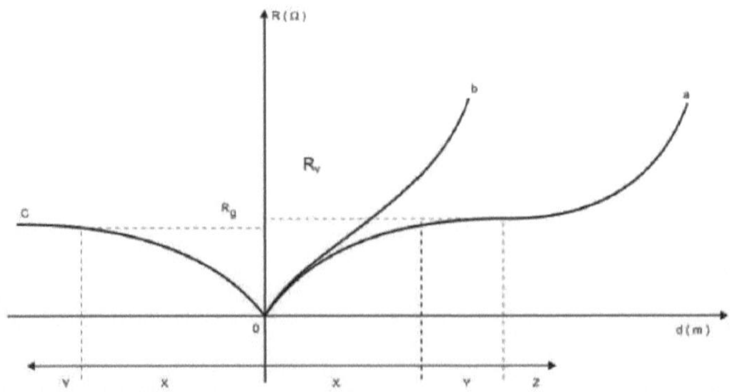

Figure 13: Typical earthing resistance curves as a function of the relative positions of the auxiliary

potential and current electrodes.

Caption
X: Area of influence of the earthing system under measurement E
Y: Potential landing zone
Z: Area of influence of the auxiliary current electrode H
RV: Grounding resistance of the system under measurement (true value of the grounding resistance).
system grounding E)
a, b, c: Earthing resistance curves as a function of the spacing and relative position of the auxiliary potential and current
electrodes.

Source: (ABNT NBR-15749, 2009)

Curves "a" and "b", presented in Figure 13, show the configuration of the

result when the displacement of the potential electrode coincided with the direction of

the current electrode. In curve "c", the direction of the potential electrode was opposite

to that of the current electrode, but the direction is the same. In this analysis, it was also

possible to see that in curves "a" and "c", assuming that

28

that the spacing between the electrodes is satisfactory, represents the plateau that corresponds to the value of the grounding resistance of the mesh under test. In the case of curve "b", the current electrode is at an insufficient distance, and the zones of influence of the earthing system and the current electrode may be overlapping, making it impossible to obtain a reliable value of the earthing resistance. For this reason, in order to make the test viable, the current electrode needs to be further away than the one used for the measurement. This procedure can also be used to obtain the "a" and "c" curves, depending on the site conditions.

Normally, the distance "d" from the periphery of the earthing system under test to the current electrode should be at least three times the largest dimension of the mesh. To check the horizontal section of the curve (the level at which the grounding resistance value should be taken), a number of measurements should be taken, varying the position of the potential electrode by 5% of "d" to the right (Sl) and to the left (S2) of the initial measurement point S. If this point is sufficient to fulfil the condition of non-overlapping areas of influence and the percentage between the difference between the values measured with the potential electrode at S1 and S2 and the value measured at S does not exceed 10%, we can take the resistance value measured at S as a correct measurement.

An important factor in avoiding measurement errors is to check for external influences in the places where the current electrode will be positioned, so that there are no buried metal conductors, metal pipes, etc. between the electrode and the earthing mesh to be tested.

With regard to the direction of movement of the potential electrode, there are advantages and disadvantages to both forms of measurement. In theory, moving the potential electrode in the same direction as the current electrode gives the true value of the resistance of the earthing system under test for a given point S. So, as a reference, for homogeneous soils, earthing systems considered to be small (largest dimension less than 10 metres) with an adequate distance "d" between the system (point E) and the current electrode (point H), point S is approximately 62% of the distance "d" away from E.

As mentioned above, moving the potential electrode S in the opposite direction to the current electrode H theoretically presents a lower resistance value than the real one, known as the lower limit of real resistance. However, this value can be perfectly acceptable, as long as H is adequately distanced from E, since the resistance value within the overall procedures for analysing earthing systems can be taken into account. The great advantage of this procedure is that the coupling effects between the current and potential circuits are minimised, and in many situations this is the only practical alternative to the potential drop method. For larger systems (dimensions greater than 10 metres), this procedure is not recommended, so to minimise errors, measurements should be taken with the current and potential electrodes aligned and in the same direction.

An important factor in grounding resistance measurements with very low values is the effect of coupling between the cables interconnecting the current and potential circuits, which is indispensable for large systems as it requires long cable lengths. Using the 60 Hz range, the errors made in the measurements are quite considerable, taking into account the inductive coupling between the parallel cables $(0.1\Omega/100m)$. As a rule of thumb, coupling problems are:

- Negligible in grounding resistance measurements above 10 Ω;

- Important for measurements below 1.0 Ω;

- Can be analysed on a case-by-case basis for measurements between 1.0 Ω and 10 Ω;

In many measurements it is essential to increase the test current in order to obtain a safe value. The simplest way when using equipment that does not allow current variation is to reduce the grounding resistance of the test electrode.

This can be done by reducing the resistivity of the auxiliary electrode installation point and/or increasing the number of rods in parallel and/or using longer rods. The maximum permissible value of the grounding resistance of each auxiliary electrode is generally specified by the manufacturers of the measuring instruments. The grounding resistance of the current electrode is usually less than 500Ω.

One of the factors that can seriously interfere with measurements made with direct current instruments are galvanic potentials, polarisation and parasitic direct currents. In general, instruments work on alternating current (not fundamentally sinusoidal). The conventional earth meter (alternating current) can be affected by parasitic alternating currents circulating in the ground, in the earthing system or in the test circuits. The way to minimise this problem is to use a test frequency that differs from the eddy currents, which is done using instruments that allow the frequency of the applied voltage to be varied, which is more suitable for these cases. A viable alternative is to use narrowband instruments and/or filters. When using equipment that operates at a higher frequency than industrial current, it is advisable to use a band filter and a synchronous rectification system, using the equation below:

$$F = \frac{(2n + 1)}{2} \cdot f \qquad\qquad (3.1)$$

Caption

F: Instrument frequency [Hz];

f: Industrial frequency [Hz];

n: Integer.

The fulfilment of this equation means that the operating frequency does not coincide with any harmonic of the industrial frequency. By using suitable filters, this makes it possible to eliminate the effect of eddy currents present in the areas studied.

This method has some limitations in its application, in certain situations it becomes very difficult or even impossible to carry out this method, such as:

- Urban installations in densely populated regions;

- Large grounding systems.

3.1.2 Potential drop method with high current injection

This method consists of circulating a high current between the earthing system under test and the ground by means of an auxiliary current electrode, measuring the potentials on its surface and obtaining the value of the earth's ohmic resistance. This method is recommended for measuring potentials on the ground surface and also the resistance of a particular earthing system or the impedance of a global earthing system, which may also involve substations with transmission line lightning cables, feeder neutral conductors, among others.

Usually a transmission line tower with a section of the line or an earthing grid from an adjacent substation or an auxiliary earthing grid built solely for this purpose is used as the auxiliary current electrode. In order to obtain a high current, it is important that the current electrode has a low resistance that is compatible with the system being tested.

In order to avoid mesh influence regions, the current electrode must be at least five times the longest dimension of the earthing system under test. This distance depends on the configuration of the earthing system, the electrical system and the type of soil. A practical way to carry out this type of measurement on systems that are not yet energised is to inject the test current into the phase conductors of the transmission lines belonging to the installation, short-circuited and interconnected to a grounded tower that is positioned more than 5 km from the system to be measured.

The potential electrode will be a metal rod firmly embedded in the ground. This electrode (S) must be moved radially from the periphery of the earthing system under test (E), and the voltage reading between S and E will be taken with a high-impedance input voltmeter. The displacement of the potential electrode must be in a direction that makes an angle between 90° and 180° in relation to the direction of the current electrode in order to avoid possible coupling between these two circuits.

The resistance of the earthing system under measurement is given by:

$$R = \frac{V}{I} \tag{3.2}$$

For the measurement result, the greater reliability of the values obtained is directly proportional to the value of the test current, as there will be less relative influence from possible interference currents. The current value will depend on the current injection source used. It is important to emphasise the safety issues relating to the personnel involved in the measurements, i.e. the higher the current, the more dangerous it becomes.

This method is particularly applicable to the earthing loops of substations, power stations and others. In this way, the current is injected exclusively between the mesh to be measured and the auxiliary current electrode, disconnecting the other alternative return paths. But in many cases there is an interest in checking the behaviour of the earthing system as a whole, including the mesh, the lightning cables, the transmission line, the neutral of the feeders, the cable shielding and others that remain when the system operates under normal conditions. The only way to perform this check is to carry out the high current injection tests with all the return paths connected to the grid, using a transmission line as a current circuit, as shown in Figure 14. This figure shows how the method can use the tool for the current electrode, which is a 2ª substation or an earthed transmission tower with the respective minimum distances of 20 km or 5 km.

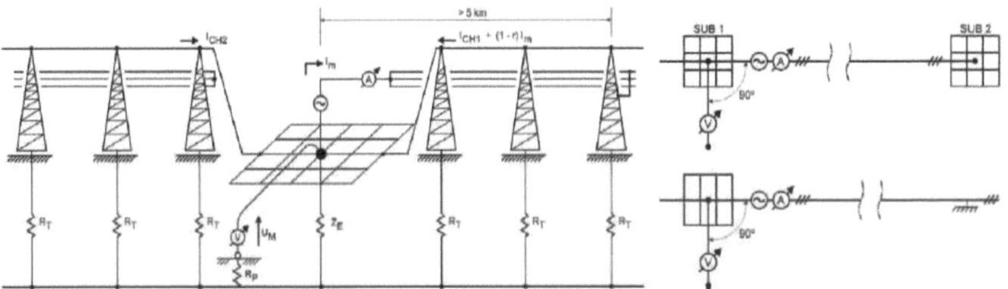

Figure 14: Large current injection method.
Source: (Adapted from ABNT NBR-15749, 2009)

The currents returning through the various paths of the interconnected earthing system can be determined directly by effective current ammeters, but the current returning through the mesh cannot be determined directly. An applicable procedure is to install a differential measuring circuit, using current transformers in the various return paths, as shown in Figure 15.

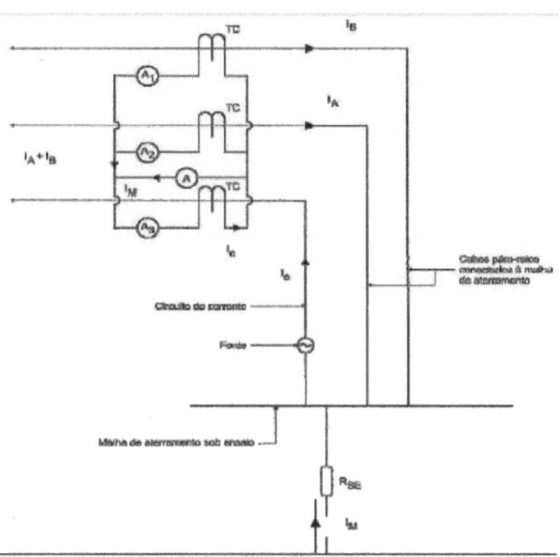

Figure 15: Monitoring the loop current.
Caption
I_e :Test current (A)
I_M: Mesh current (A)
I_A ,I_B : Currents through lightning cables (A)
R_{SE}:   **Grounding resistance of the installation (fi)**

Source: (Adapted from ABNT NBR-15749, 2009)

3.1.3 Pliers terrometer

Due to its practicality, ease of transport and handling, measuring with this equipment has become more popular and is proving to be very efficient in determining the grounding resistance in the systems that serve electrical installations, especially in densely built-up areas.

The working principle of a pliers terrometer is based on an alternating current generator that applies a voltage to a coil with N turns, the ferromagnetic core of which surrounds a closed circuit, which corresponds to a

a single secondary coil of a transformer with an N:1 ratio. A voltage is then applied to the coil which produces a known electromotive force (e.m.f). With another coil of M turns, the current travelling through the circuit is measured, which can be seen in Figure 16.

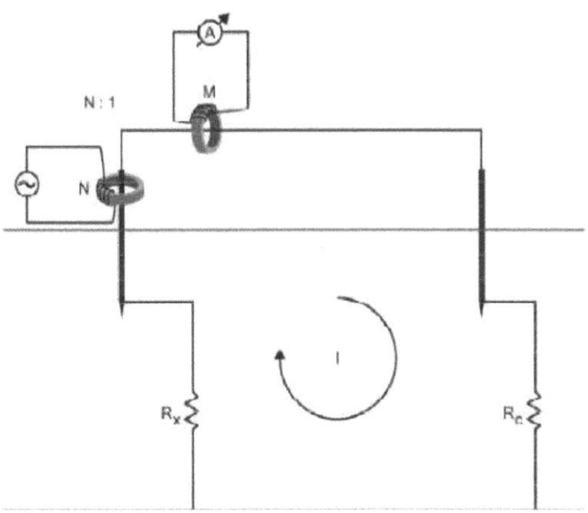

Figure 16: Working principle of a pliers terrometer.
Source: (Adapted from ABNT NBR-15749, 2009)

The sum of the resistances Rx + R_c gives the ratio between the voltage generated and the current circulating in the circuit. As Rc represents a set of electrodes in parallel, Rx can be considered much larger than $_{Rc}$. With this condition, the pliers will measure the value of the grounding resistance $_{Rx}$. The corresponding electrical circuit can be seen in Figure 17 below.

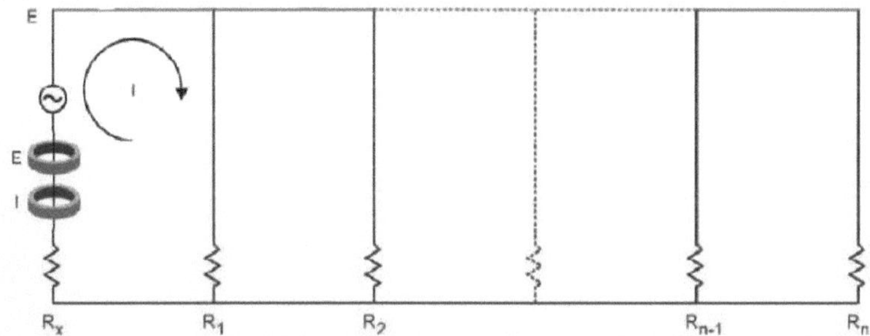

Figure 17: Electrical circuit.
Source: (Adapted from ABNT NBR-15749, 2009)

The set of electrodes R1 to Rn are replacing Rc, so when a voltage E is applied to electrode Rx via a special transformer on the solenoid, a current I circulates through the circuit, we can deduce the following equation:

$$\frac{E}{I} = R_x + \cfrac{1}{\displaystyle\sum_{k=1}^{n} \frac{1}{R_k}} \tag{3.3}$$

With the condition below, the final resistance is

$$R_x \gg \frac{1}{\displaystyle\sum_{k=1}^{n} \frac{1}{R_k}} \qquad\qquad R_x = \frac{E}{I} \qquad\qquad (3.4)$$

These meters are built in the form of a pair of split-core pliers and are large enough to wrap around the conductors of the earthing system.

One of the cores generates an electromotive force (e.m.f.), which in turn produces the electric current circulating in the test circuit, and the other is a current measurement transformer. In order to attenuate disturbances caused by the presence of spurious voltages, which would produce errors in the results obtained or even make it impossible to carry out the test, the equipment generally works with different measuring frequencies (between 1.5 kHz and 2.5 kHz) from the industrial frequency and has suitable filters.

The great advantage is that there is no need to drive auxiliary rods into the ground and the reduction in the number of conductors used creates a natural tendency to test all earthing systems with a pliers terrometer. For this use, at least the following requirements must be met:

- The method can be applied to measuring earth resistance when there is a closed circuit (loop), including the earth resistance to be measured;

- The resistance of the earthing system that closes the loop must be much lower than the resistance of the earth under measurement ($R_c + R_{cable} \ll R_x$);

- The distance between the ground under measurement and the nearest of the grounds that close the loop must be large enough so that the respective zones of influence do not overlap;

- The resistance of the system under measurement must be covered by the entire current injected into the ground, i.e. the positioning of the equipment and the auxiliary conductor is extremely important for the test.

3.2 Measuring potentials on the soil surface

The methodology for measuring potentials on the soil surface is similar to that used for measuring electrode resistance values. It is advisable that the

The potential profiles on the surface of the ground and the measurements of touch and step voltages are carried out by injecting high values of electric current into the ground. For these measurements, voltmeters and ammeters with scales suitable for the measurement ranges must be used, or a dedicated instrument that fulfils the necessary conditions. For measurements in certain locations of earthing systems such as houses, simple buildings and places where there is no suspicion of strong eddy currents, a common earth meter can be used to initialise the measurement, always following the manufacturer's specifications.

The procedures described aim to determine the surface potentials due to industrial frequency currents, mainly short-circuit currents, with the main objective of ensuring the safety of people travelling on and around earthing systems. The potentials that occur in earthing systems due to high-frequency currents, such as lightning strikes, must be treated specifically.

The current injection circuit must be set up in a similar way to the earth resistance measurement. The current values that will be injected into the earthing loop must have values that are compatible with the measurement system.

Potential measurements must be taken at points previously marked out in the project or in the measurement planning, in strategic areas of the substations, using a voltmeter with a high input impedance, generally not less than 1 $M\Omega/V$. Electronic voltmeters are generally suitable for this purpose.

3.2.1 Touch voltage measurement

The touch voltage must be measured between metallic elements such as metallic structures, equipment housings and metallic masses connected to the earthing system under study and the potential electrode embedded in the ground as shown in Figure 18 or as indicated in Figure 19 (use of plates as auxiliary electrodes), always observing a distance of 1 metre.

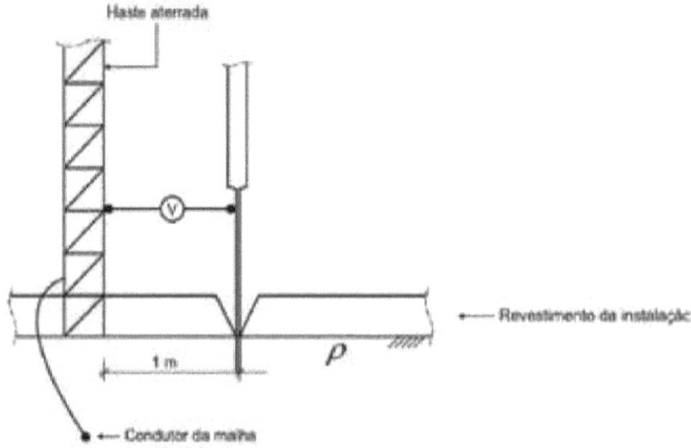

Figure 18: Touch potential measurement.
Source: (Adapted from ABNT NBR-15749, 2009)

3.2.2 Step voltage measurement

The step voltage must be measured between two potential electrodes embedded in the ground, 1 metre apart, as shown in Figure 18 or as indicated in Figure 19 (use of plates as auxiliary electrodes).

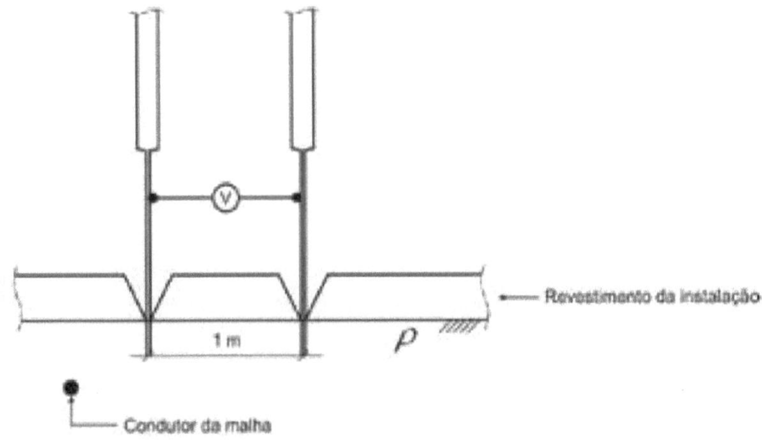

Figure 19: Step potential measurement.
Source: (Adapted from ABNT NBR-15749, 2009)

The electric current source used to carry out the measurements must have adequate power and voltage to provide sufficient electric current values. These values are generally high in order to reduce measurement errors due to the interference currents that generally circulate in the ground. The source used can be a motor-generator set or an isolating transformer (step-down or not) connected to the power grid near the measurement site. It is recommended that a power supply with an adjustable output voltage is used.

The ABNT NBR 15749:2009 standard also presents some methods and alternatives that help with measurements, such as the synchronous method at industrial frequency, capacitive compensation and the beat method. This is an alternative to increasing the injected current in order to minimise the effect of eddy currents or in cases where there are significant interference currents in relation to the test current.

During measurements, the lightning cables and counterweights of the transmission lines, the neutrals of the transformers, the armouring and the metal covers of insulated cables arriving at the installation must be disconnected from the earthing system under test.

The value of the test current is very important for this measurement, because the higher the current injected, the higher the measured voltage values, and therefore the lower the relative influence of interference currents. On the other hand, in order to obtain these higher currents, the source of electric current injection is limited, increasing the safety problems for the personnel involved in the measurements and those who may be in the vicinity. Generator/source test voltages of around 100 V are normally used in these measurements.

A very important aspect to consider is the choice of preferred locations for measuring potentials on the soil surface, such as:

- Ideally, a complete mapping of the installation should be made and measurements taken to cover the entire area to be investigated. However, this is often not done, usually due to time constraints, and measurements should preferably be taken on the periphery of the earthing system, where the highest voltages are generally found on the ground surface, and some measurements should be taken in the central region of the system (there is the possibility of the presence of people or non-uniform spacing).

- With regard to measuring the touch voltage on grounded metal parts, it is often not known where the highest voltage is obtained, so it is recommended to take several measurements (at least 3) in different directions (particularly the directions away from the buried grounding conductors and/or those towards the periphery of the system).

- For the test with the entire earthing system interconnected, currents of the order of 100 A or more are generally required for the values obtained for the voltage at the ground surface to be reliable.

The measured voltage values must be corrected for the actual value of the loop current during an earth fault, established for the worst fault condition. This correction must be made according to the following equation:

$$V_R = \frac{V_e \cdot I_M}{I_e} \tag{3.5}$$

Caption

V_R :Actual voltage during an earth fault [V];

V_e :Voltage measured during the test [V];

I_M: Mesh current [A];

I_e :Test current [A].

The foot-gravel (or ground) contact resistance can be determined in the current injection test, as can the voltage applied directly to the person. Figure 20 provides details for measuring touch and step voltages and shows the equivalent circuits.

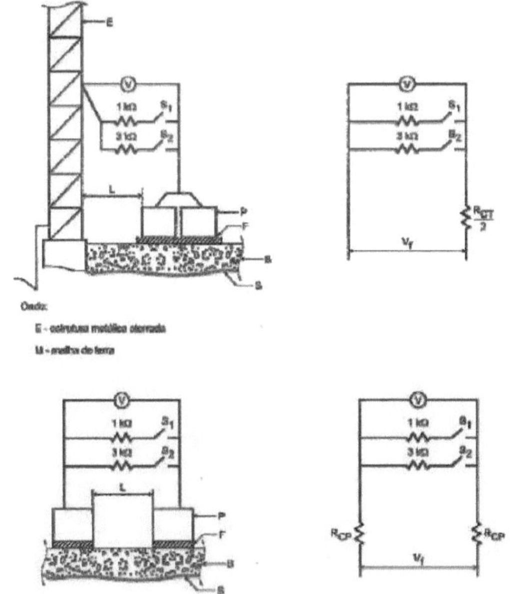

Figure 20: Measurement of touch and step voltages.
Caption
E: Grounded metal structure
S: Soil
F: Felt soaked in water and salt
L: 1.0 m
B: Crushed stone (gravel)
P: Weight 25kg
V: Electronic voltmeter

Source: (Adapted from ABNT NBR-15749, 2009)

CHAPTER 4

CASE STUDY

The substation is the part of the power system that includes the manoeuvring, control, protection and transformation devices and other equipment, conductors and accessories. Therefore, for a company to function perfectly, it must have a properly functioning system, protecting the equipment and, above all, the operators.

The case study was carried out in an aerial substation with its equipment subject to inclement weather, in a textile industry with a supply voltage of 69kV and an installed power of 5/6.25MVA. Grounding resistance measurements were carried out using the potential drop method, and ground surface potential measurements using a method similar to the resistance method by applying a high current and injecting it through rods.

The design of the earthing grid for this substation is shown below in figure 21 and in the attached drawing.

Therefore, the case study was an analysis of the results measured at the substation, such as the values of grounding resistance and touch and step voltage, to check whether the values are satisfactory, being lower than the maximum values stipulated by Coelce's NT-004/2011 R05 (High voltage electricity supply - 69kV) standards and IEEE Std 80-2000: "IEEE *Guide for Safety in AC Substation Grounding*".

The earthing mesh is designed to meet the requirements of mechanical resistivity, good condition and capacity to withstand short circuits. The purpose of this mesh is to ensure that:

- Ensure that people located in and around the substation are not exposed to the danger of a critical electric shock, with regard to touch and step potential limits;
- To provide an effective means for electrical currents to flow to earth, under normal or fault conditions, without exceeding any equipment operating limits, as well as allowing the protection to function perfectly;

- Obtain the lowest possible earth resistance for earth fault currents.

The results of the tests will be presented below using the values of the measurements for the substation's earthing system, considering the value of the resistance and the potentials on the ground surface, which will verify the performance of the design of the earthing mesh to be inspected, respecting the maximum values of earthing resistance and touch and step potentials as a function of the designed mesh.

Data Considered for the Ground Mesh

- Apparent soil resistivity: $\rho a = 336$ $\Omega.m$
- Resistivity of the first layer: $\rho l = 207.33$ $\Omega.m$
- Thickness of the first layer: $a1 = 0.89m$
- Resistivity of the second layer: $\rho 2 = 840$ $\Omega.m$
- Surface resistivity of gravel: $\rho s = 3,000 \Omega.m$
- Phase-to-earth short-circuit current: $Icft = 11,487.45$ A
- Mesh length: $Cm = 50$ m
- Mesh width: $Lm = 12$ m
- Distance between main and junction conductors: $Dcp = 2m$ $Dcj = 2m$
- Current circulation time through the mesh: $Te = 3s$
- Protection operating time: $= 0.5s$

  -Conductor diameter : $d = 0.00975$ m

  -Mesh depth: $h = 0.5m$

  -Human body resistance : $Rch = 1.000 \Omega$

  -Ground rod diameter : $Dht = 3/4$ inch

  -Length of ground rod: $Lh = 6m$

-Number of earth rods used initially: $Nh = 21$

  -Gravel thickness : $hs = 0.1$ m

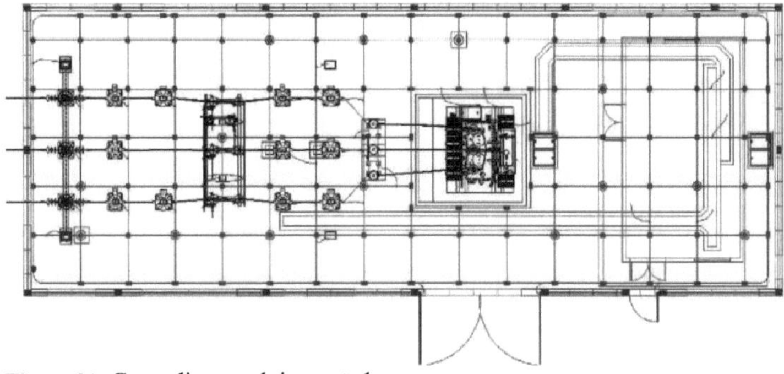

Figure 21: Grounding mesh inspected.
Source: own elaboration, 2013.

4.1 Grounding Resistance Measurements

Nature of Service: Grounding Mesh Inspection

4.1.1 Measurement sketch:

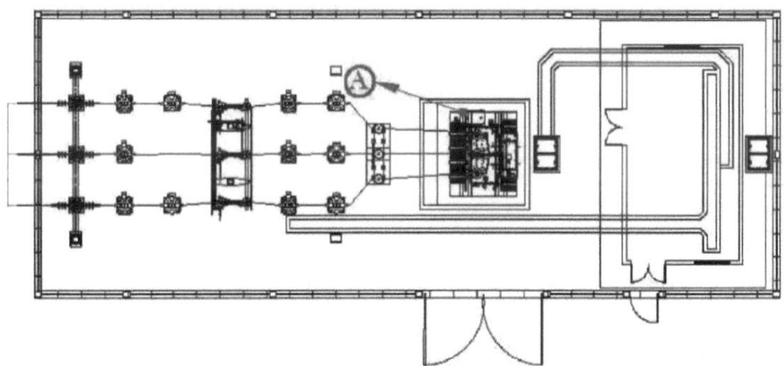

Figure 22: Grounding resistance measurement points.
Source: own elaboration, 2013.

4.1.2 Measurement results:

Table 4: Grounding resistance measurement values.

LOCAL	POINT	LOCAL	POINT
Transformer	A	Transformer	A
Distance (m)	Values Measured (Ω)	Distance (m)	Values Measured (Ω)
0	o,n	315	2,99
5	0,19	325	2,99
10	0,23	335	2,99
15	0,3	345	2,99
20	0,39	355	3,05
25	0,55	365	3,15
30	0,65	375	3,25

35	0,75	385	3,4
40	0,85	395	3,65
45	0,95	405	3,95
55	1,15	415	4,25
65	1,35	425	4,6
75	1,65	435	5,1
85	1,8	445	5,5
95	1,95	455	5,9
105	2,15	465	6,1
115	2,35	475	6,8
125	2,45	485	7
135	2,5	495	8,5
145	2,6	500	9
155	2,68		
165	2,73		
175	2,75		
185	2,77		
195	2,79		
205	2,81		
215	2,85	-	-
225	2,88		
235	2,9	-	-
245	2,93		
255	2,96		
265	2,97		
275	2,98		
285	2,99		
295	2,99		
305	2,99	-	-

Source: own elaboration, 2013.

Figure 23 below shows the displacement of the potential electrode to generate the mesh resistance curve.

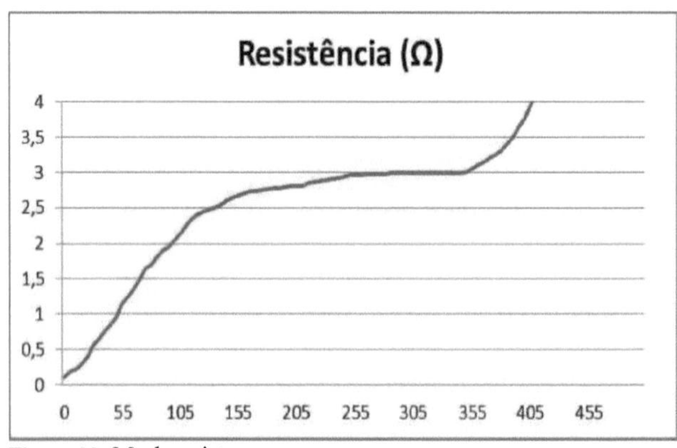

Figure 23: Mesh resistance curve.

4.1.3 Measurement parameters:

-Instrument   Used: Digital Terrometer; brand: Megabrás, model: MTD- 20 kWe

 -Average spacing of   auxiliary measuring electrodes:

- Current electrode: 500 m
- Potential electrode: 310 m

- Ambient temperature: 32 °C
- Soil conditions: Dry
- Measurement date: 11/05/2013

4.1.4 Conclusions:

The values obtained are satisfactory when compared with the requirements of the Coelce Standard **NT-004/2011 R-05 "High voltage electricity supply - 69 kV",** which stipulates a maximum value of **5 Ω** for the resistance of the earth mesh. Substituting the values from the equation cited in chapter two, we arrive at a maximum grounding resistance of 2.99 Ω.

4.2 Touch Voltage Measurements

Nature of Service: Grounding Mesh Inspection

An alternating current of 100 A was injected into points 1 and 2 of the earthing grid, and measurements were taken between the metal casing of the equipment and an electrode spaced one metre away from it, at points A, B, C, D, E, Fe G. The following table shows the results of these measurements and their extrapolated values for the (maximum) phase-to-earth short-circuit current of 11,487.45 A. No measurements were taken on the periphery of the grid, as all the points were requested by the local utility company.

4.2.3 Measurement sketch:

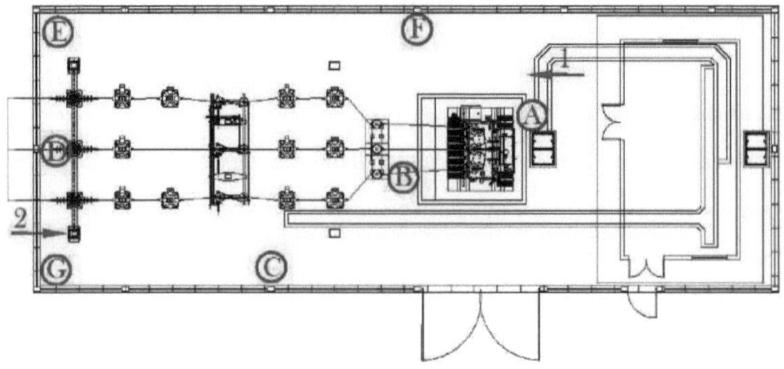

Figure 24: Touch voltage measurement points.
Source: own elaboration, 2013.

4.2.   2Measurement results:

Table 5: Touch voltage measurement values.

POINT	MEASURED VALUES	EXTRAPOLATED VALUES
A	47.0 DIV$_{AC}$	5.40 V$_{AC}$
B	47.0 mV$_{AC}$	5.40 V$_{AC}$
C	75.0 mV$_{AC}$	8.62 V$_{AC}$
D	590.0 mV$_{AC}$	67.78 V$_{AC}$
E	640.0 mV$_{AC}$	73.52 V$_{AC}$
F	620.0 mV$_{AC}$	71.22 V$_{AC}$
G	47.0 mVAC	5.40 V$_{AC}$

Source: own elaboration, 2013.

4.2.   3Measurement parameters:

-Instruments   Used: Current Source.

Digital Multimeter, brand: Fluke , model: 87V

Digital clamp meter, brand: Fluke , model: 336

-Injected current   : 100 A

-Earth-phase short circuit   : 11,487.45 A

-Average resistivity of gravel (ps): 3,000 Ωm

-Short-circuit current duration time   (tc): 0.5 s

-Ambient temperature   : 32 °C

-Soil Conditions   : Dry

-Measurement date   : 11/05/2013

4.2.4 Conclusions :

According to **IEEE Std 80-2000:** ***"IEEE Guide for Safety in AC Substation Grounding***', the maximum touch voltage is given by the following equation:

$$E_{toque} = (1000 + 1,5.Cs.\rho s).\frac{k}{\sqrt{Te}} \qquad (4.1)$$

Substituting the values from the equation in chapter two, we arrive at a maximum touch voltage of 689.00 V.

1.3 Step Voltage Measurements

Nature of Service: Grounding Mesh Inspection

An alternating current of 100 A was injected into points 1 and 2 of the earthing grid, and measurements were taken between two electrodes spaced one metre apart, at points A, B, C, D and E. The following table shows the results of these measurements and their extrapolated values for the (maximum) phase-to-ground short-circuit current of 11,487.45 A. No measurements were taken on the periphery of the grid, as all the points were requested by the local utility company.

4.3.   1 Measurement reports:

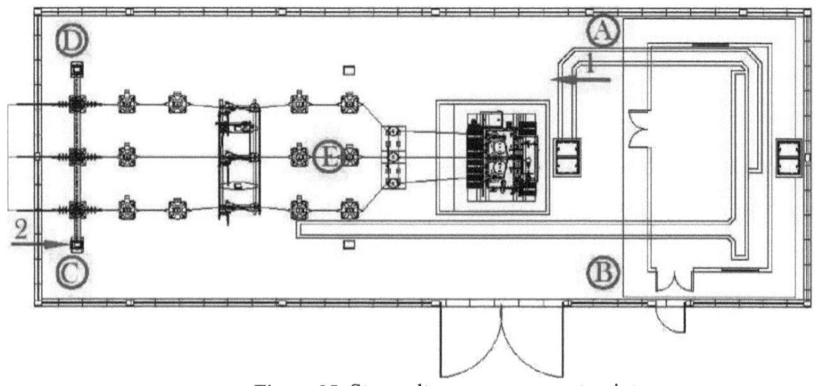

Figure 25: Step voltage measurement points.

Source: own elaboration, 2013.

4.3.2 Measurement results:

Table 6: Step voltage measurement values

POINT	MEASURED VALUES	EXTRAPOLATED VALUES
A	48.0 diV$_{AC}$	5.51 V$_{AC}$
B	122.4 diV$_{AC}$	14.06 V$_{AC}$
C	67 mV$_{AC}$	7.69 V$_{AC}$
D	140.0 mV$_{AC}$	16.08 V$_{AC}$
E	82.0 mVAC	9.42 V$_{AC}$

Source: own elaboration, 2013.

4.3.3 Measurement parameters:

- Instruments Used: Current Source.

  Digital Multimeter, brand: Fluke, model: 87V

  Digital clamp meter, brand: Fluke, model: 336

- Injected current: 100 A

- Phase-ground short-circuit: 11,487.45 A

- **Average gravel resistivity (ps): 3,000 Ωm**

- Duration of the short-circuit current (tc): 0.5 s

- Ambient temperature: 32 °C

- Soil conditions: Dry

50

- Measurement date: 11/05/2013

4.3.4 Conclusions:

According to **IEEE Std 80-2000:** *"IEEE Guide for Safety in AC Substation Grounding'* ', the maximum step voltage is given by the following equation:

$$E_{passo} = (1000+6.Cs.\rho s).\frac{k}{\sqrt{Te}} \qquad (4.2)$$

By substituting the values from the equation in chapter two, we arrive at a maximum step voltage of 2,263.85 V.

1.4 Final opinion on the installation

The results obtained from the inspection of the substation's earthing grid indicate that the earthing system is in satisfactory condition, as presented in the project.

The visual inspection carried out shows that the installation is in a good state of cleanliness and there is no corrosion on the equipment.

It was also checked that all the substation's equipment and components are properly grounded, such as fittings, metal doors and screens, cable armouring, protective conductors, as well as the transformer's neutral.

CHAPTER 5

CONCLUSION

The comparative analyses between the results obtained with the criteria used in each method and the values estimated by the standards indicate whether the earthing system is effective, fulfilling its role, seeking safety measures and aiming to guarantee the protection of facility personnel. If the system is not satisfactory, changes should be made to avoid accidents.

Considering that companies depend directly on electricity to run their production processes, we can see that the substation is an extremely important element for these consumers. Therefore, its proper functioning is essential.

The electricity supply is released to the customer once the standards prescribed by Coelce have been met, with the aim of ensuring compliance with the project and the perfect installation of the system. Therefore, one of the tests of this commissioning is that all equipment must be grounded and have an adequate grounding system.

If the results of the final verification are satisfactory for all requirements, the substation will be authorised to be energised.

As a result of this work, it was possible to understand the importance of inspecting the substation grounding grid before energising it, as it is a very useful tool for detecting possible problems in system operation.

The purpose of the study, then, was to present the concepts of inspecting a grounding grid in a 69/13.8kV substation, which is already covered by current standards, such as IEEE Std 81-2012: "IEEE *Guide for Measuring Earth Resistivity, Ground Impedance, and Earth Surface Potentials of a Ground"*.
System" and ABNT NBR-15749-2009 "Measurement of earthing resistance and potentials on the ground surface in earthing systems", introducing a specific measurement methodology for each element, such as resistance and potentials.

Therefore, it was concluded that the substation is within the suitable values of the earth grid. This work has shown that all the regulatory requirements have been met and that, in the event of faults, the system will guarantee the protection of operators, who will be exposed to the minimum risk of accidents related to potential hazards that may occur in the vicinity of the earthing system.

It can be concluded that measurements on the earthing system can be made in various ways, as mentioned in this paper, without compromising the final results or their accuracy, but always following the recommendations of the standards and the limitations of each method used for its purpose.

In addition, it demonstrates the importance of an initial and periodic inspection of the substation earthing system, clearly showing the need for an adequate system, preventing and ensuring safety.

CHAPTER 6

BIBLIOGRAPHY

Brazilian Association of Technical Standards. NBR 15749: Measurement of Grounding Resistance and Potentials on the Ground Surface in Grounding Systems - Elaboration. Rio de Janeiro. 2009.

Brazilian Association of Technical Standards. NBR 6023: Information and documentation - References - Elaboration. Rio de Janeiro. 2002.

Brazilian Association of Technical Standards. NBR 14724: Information and documentation - Academic papers - Presentation. Rio de Janeiro. 2011.

COELHO, ROONEY RIBEIRO. 2012. Electromagnetic Analysis of Grounding Systems in Quasi-Static Regime Using the Finite Element Method. 1. Fortaleza : Unifor, 2012.

Companhia Energética do Ceará. NT-004/2011 R-05: High voltage electricity supply - 69kV. Fortaleza. 2011.

KINDERMANN, GERALDO and CAMPANOLO, JORGE MARIO. 2011. *Electrical Earthing.* 6. Santa Catarina : EEL, 2011.

KINDERMANN, GERALDO. 2005. *Electric Shock.* Santa Catarina : EEL, 2005.

LEITE, CARLOS MOREIRA. 2007. *Earth Mesh: Electrical Earthing Techniques.* 1. São Porto : Editora Oficina de Mydia, 2007.

MAMEDE FILHO, JOÃO. 2010. *Industrial Electrical Installations.* 8. Rio de Janeiro : LTC, 2010.

TELLÓ, MARCOS. 2007. *Impulsive Electrical Grounding at Low and High Frequencies.* 1. Porto Alegre : ediPUCRS, 2007.

The Institute of Electrical and Electronics Engineers. IEEE std. 80: Guide for Safety in AC Substation Grounding. 2000.

The Institute of Electrical and Electronics Engineers. IEEE std. 81: Guide for Measuring Earth

Resistivity, Ground Impedance, and Earth Surface Potentials of a Ground System. 2012.

VISACRO FILHO, SILVERIO. 2002. *Electrical Grounding - Basic Concepts, Measurement and Instrumentation Techniques, Grounding Philosophies.* 1. São Paulo : Artliber, 2002.